HEALTH RESEARCH
Explainers

Foundations

PAUL IAN CROSS, PhD

Text © Paul Ian Cross, 2024

First edition published in Great Britain in 2024.

Farrow Books

London

United Kingdom

Cover designed by James at GoOnWrite.com

British Library Cataloguing in Publication data available.

ISBN: 978-1-912199-39-6

e-ISBN: 978-1-912199-31-0

Contents

References

Health Research Explainers is dedicated to the participants who take part in health research.

On behalf of the health research community, we extend our deepest gratitude for your involvement. Together, we can build a healthier future for all.

Introduction

As a scientist and researcher with over two decades of experience in health research, I am grateful to have been involved in setting up and delivering hundreds of clinical studies covering multiple disease areas.

My first role was working for an early-phase clinical research unit where I worked on first-in-human clinical trials with healthy volunteers. Next, I moved to the NHS where I coordinated multiple studies in pain and stroke research.

Later, I worked at Great Ormond Street Hospital for Children managing clinical trials in rare diseases, oncology and epilepsy. At Guy's and St Thomas' NHS Foundation Trust I worked on studies in asthma and allergy, oncology, gastroenterology and gene and cell therapy.

Throughout my career, I am proud to have been involved in several novel studies. These advancements were made possible through science and the dedicated efforts of researchers, healthcare professionals and study participants.

However, over the years, I have also noticed a significant gap in understanding between the scientific community and the public regarding research principles and practices. This issue became apparent during the COVID-19 pandemic when the spread of misinformation contributed to a lack of trust in science and scientists.

This inspired me to write this book, aimed at early-stage researchers, patients and the public. In the book, I explain complex science in easy-to-understand language.

As someone who is passionate about education and making science accessible to all, I hope this book will promote informed decision-making and encourage participation in clinical studies. And of course, ultimately, drive progress in healthcare.

I hope this book will not only inform but also inspire you to become an active participant in the pursuit of better health for all. I am excited to share my passion for science and research with you.

Here's to the future,

Paul

Paul Ian Cross, BSc (Hons), MSc, PhD

Chapter One

What is Health Research?

H ealth research, also known as medical research, is dedicated to investigating and improving various aspects of health, including patient care, disease prevention and treatment. The goals of health research are to improve healthcare outcomes, enhance patient well-being and shape health and care policies[1].

In simple terms, health outcomes are the results or effects of healthcare on a person's health. They measure how well a treatment, medical intervention or care improves things like recovery from illness, quality of life or overall well-being. Examples include feeling better after surgery, managing a chronic condition or living longer due to medical treatment.

Health outcomes help doctors and researchers understand what works best to keep people healthy[2].

Health research often begins with unanswered questions. These questions drive the formulation of hypotheses and the design of studies to test these hypotheses. Ever wondered how drugs like aspirin are studied? Researchers run studies to explore various aspects of health and disease. For example, with aspirin, researchers might study how it works to reduce pain. This could include looking at how aspirin affects the body's pain sensors, how it helps reduce swelling and how it works for different kinds of pain.

Health research may involve testing new medicines, vaccines, surgical techniques or medical devices, often involving both known and unknown risks. Researchers generate research questions. To make sure they can answer these questions, they follow detailed plans known as protocols.

A research protocol outlines the objectives, methodology and procedures of a study. Protocols are designed to protect the safety of participants, make sure the data is accurate and ensure the reproducibility of the results — this means that the results can be repeated. They include detailed information on the study design, participant recruitment, safety monitoring, data collection methods and statistical analysis plans.

> *Dr Paul Science Says*
> 'Health research involves asking questions about our health and well-being. But the difficult part is asking the right questions.'

Several landmark studies have significantly advanced our understanding of health and disease. Here are some examples:

The Framingham Heart Study (Ongoing since 1948)

The Framingham Heart Study is a long-term, on-going research project focused on cardiovascular health, involving residents of Framingham, Massachusetts. It started in 1948 with 5,209 adult participants and is now in its third generation of subjects. Before this study, very little was known about the causes and patterns of high blood pressure and cardiovascular disease.

Many of the widely accepted insights into heart health, such as the impact of diet, exercise and medications like aspirin, come from this study. It is conducted by the National Heart, Lung and Blood Institute in partnership with Boston University since 1971, with healthcare professionals from the Greater Boston area supporting the research[3].

The Human Genome Project (1990-2003)

An organism's genetic information is called its genome[4]. The Human Genome Project (HGP), launched in the late 20th century, was developed to identify all the genes in the human genome and the sequence of all three billion base pairs. The project was completed in 2003, but work continues to identify all human genes. The HGP used DNA from multiple individuals to produce an 'average' genome sequence, though each person has a unique genome (except identical twins).

There is still much to learn about how knowledge of the human genome can advance medical treatments. Scientists hope that information from the HGP will lead to new methods for diagnosing and treating diseases, especially genetic disorders and cancer.

Cystic fibrosis, for example, is caused by two damaged alleles on chromosome [7]. In simple terms, an allele is a version of a gene. Genes come in pairs, and each one has different versions, or alleles, that can influence traits like eye colour or hair type.

Some alleles can be healthy, while others might cause health problems. In the future, gene therapy might be able to correct errors, potentially by inserting a healthy version into lung cells, which could greatly improve the lives of patients who often require lung transplants as the disease progresses. Understanding the genome has made gene therapy for cystic fibrosis more feasible.

Another example is in breast cancer. Some families have a higher risk of breast cancer due to faulty alleles. Following the HGP, these alleles have been identified, allowing for genetic testing to predict a person's likelihood of developing breast cancer. This enables individuals to make informed decisions, such as opting for preventive surgery. It also helps doctors assess the risk of passing the condition to future children.

Another major milestone was when researchers discovered how to sequence DNA in cancer cells. Comparing these sequences to the human genome helps identify mutated genes. Once they were able to recognise mutated genes, it was possible to develop new treatments that target them. Researchers discovered that some breast cancers are caused by changes (mutations) in a gene called HER-2.

This led to the development of a special treatment, called a monoclonal antibody, that targets only the cancer cells with this mutation. The treatment has been very successful and was made possible thanks to advances in studying DNA.

The Pfizer-BioNTech (BNT162b2) COVID-19 Vaccine Clinical Trials

The development of vaccines has always been seen as a remarkable achievement in medicine. However, the development of vaccines in response to the COVID-19 pandemic was a particularly notable example of rapid and effective health research. It highlights how collaboration, scientific innovation and dedication can lead to advancements in global health.

Vaccines work by harnessing the body's natural defences to build immunity and reduce the risk of illness. When you receive a vaccine, your immune system recognises the invading germ, such as a virus or bacteria, and responds by producing antibodies — proteins that the body naturally creates to combat disease.

Importantly, the immune system also remembers how to fight the disease in future, so if you are exposed to the germ again, it can act swiftly to eliminate it before you become ill. Vaccination is therefore a safe and effective way to trigger the body's immune response without causing serious illness[5].

In April 2020, the world was dealing with the COVID-19 pandemic. At the same time, scientists were racing to develop a vaccine. The first step was studying the SARS-CoV-2 virus to create the vaccine. Once it was developed, it needed to be tested in early phase (phase I and II) clinical trials.

These trials focused on safety and to understand if it was capable of triggering an immune response. These studies also aimed to identify the optimal dosage of the vaccine. Safety is always the primary concern in early phase trials and the participants were closely monitored to ensure that no serious side effects occurred.

At the same time, researchers analysed blood samples to see how well the vaccine prompted the body to produce antibodies. These are proteins that can neutralise the virus and prevent infection.

By July 2020, the vaccine had moved into phase 3 clinical trials involving approximately 43,000 participants worldwide. During this time, researchers studied how effective the vaccine was across a diverse population.

Large-scale trials like these are essential because they offer a broader understanding of how the vaccine works in people from different backgrounds, with varying ages, health conditions and even underlying health issues. In these trials, the vaccine showed strong protection against severe cases of the disease, helping to reduce hospitalisations and deaths[6].

These results led to the rapid approval of the vaccine around the world. Governments, regulators, healthcare organisations and communities worked together to roll out vaccination programmes at unprecedented speed. The vaccine played a pivotal role in controlling the spread of COVID-19 and allowed societies to gradually return to a sense of normality[7].

The success of the COVID-19 vaccines shows how far health research has come. The ability to develop, test and distribute a new vaccine in less than a year demonstrates the progress we have made in science, technology and global collaboration.

> *Dr Paul Science Says*
> 'Vaccines are the result of decades of research, planning and global cooperation.'

My COVID-19 Pandemic Experience

During the COVID-19 pandemic, I had a variety of roles to support healthcare while also navigating the intense challenges we all faced, both personally and professionally. While working part-time at the UK NIHR, I developed online training for NHS healthcare professionals redeployed into research roles — an important step in expanding the research workforce.

As a clinical research consultant, I was also asked to audit one of the COVID-19 vaccine trials. In addition, this time marked the beginning of my journey as a science communicator on social media. I joined various online groups of doctors, scientists and researchers from all around the world to explain

COVID-19 and the development of the vaccines to the public. Science communication has since become a significant part of my work, bridging the gap between research and the public.

Key Terminology and Concepts

To better understand health research, it's important to understand some of the key terms and concepts that researchers frequently use.

The Scientific Method

The scientific method is a step-by-step way of exploring questions and solving problems. It starts by identifying a question or problem, then making a guess (called a hypothesis) about what the answer might be. Researchers test their guess by collecting information through experiments or studies. After looking at the results, they decide whether the guess was correct or not. Based on what they find, they may ask new questions or adjust their original guess[8].

Protocol

A protocol is a detailed plan that outlines the objectives, design, methods and structure of a research study[9]. It acts as a blueprint, ensuring the study is conducted in the same way[9]. However, real-life challenges often disrupt protocols. For example, a protocol may specify a time window for taking blood samples, but unforeseen delays such as patient unavailability or equipment issues can make this impractical.

Similarly, while a drug dosage schedule may be mapped out, patients might miss doses due to side effects or personal circumstances. Follow-up visits might be missed due to transportation difficulties or clinic capacity issues.

These logistical problems are common and changes to the protocol (called amendments) are often needed to adapt to reality while maintaining participant safety and data quality. It's not uncommon for studies to have multiple protocol versions as they progress — one trial I worked on had 18 protocol amendments!

> *Dr Paul Science Says*
> 'Every research protocol is a blueprint, but like any good plan, it needs flexibility to handle real-world challenges.'

Participants and Informed Consent

Participants are the individuals who take part in a study, either receiving the treatment being tested or being part of a control group for comparison. Before enrolling, individuals must take part in the informed consent process. This involves providing potential participants with detailed information about the study, its purpose, procedures and any possible risks or benefits. Informed consent ensures that participants fully understand what they are agreeing to and can ask questions before deciding.

Test vs. Control Group

A control group is essential in determining whether a treatment is truly effective. By comparing outcomes between the test group (who receive the treatment) and the control group (who do not), researchers can evaluate whether the observed changes are due to the treatment itself or other factors. This comparison helps provide more reliable conclusions.

Bias and Confounding

Bias in research means anything that unfairly affects the study's results. It can happen because of how participants are chosen, how information is gathered or how the results are looked at. To avoid bias, researchers often use randomisation, which means assigning people to different groups by chance. This helps make sure the groups are similar, except for the treatment they get, making the results more reliable. Confounding is when something outside the study affects both the treatment and the result. This makes it hard to know what really caused the changes. By finding and controlling these outside factors, researchers can make their results more accurate and trustworthy.

Dr Paul Science Says
'Science is about progress, not perfection. Each study builds on the last, paving the way for breakthroughs.'

Blinding

Blinding is a way to reduce bias in studies. In a blinded study, the participants, the researchers, or both don't know who is getting the real treatment and who is in the control group. In a double-blind study, neither the participants nor the researchers know who is in which group. This helps make sure that their opinions or expectations don't affect the results, which is important in clinical trials where even small expectations can impact the results.

Efficacy vs. Effectiveness

Efficacy is about how well a treatment works in perfect, controlled conditions, like in a clinical trial. Effectiveness is about how well the treatment works in the real world. Sometimes, a treatment that works very well in a trial might not be as good in real life because of things like whether patients follow the treatment correctly or differences in healthcare.

Key Takeaways

- Health research aims to enhance healthcare outcomes, improve patient well-being and guide policy. It involves developing better ways to prevent and treat disease.

- Health outcomes are the measurable effects of healthcare on an individual's health, such as recovery from illness, quality of life and longevity. Health outcomes help researchers determine what works best to keep people healthy.

- Health research starts with questions and hypotheses, leading to carefully designed studies. These studies explore various aspects of health and often involve testing new treatments, vaccines and medical devices.

- A protocol is a detailed study plan that ensures safety, accuracy and reproducibility. It includes specifics on study design, participant recruitment and data collection.

- Influential studies like the Framingham Heart Study, the Human Genome Project and the COVID-19 vaccine trials showcase the impact of health research in advancing knowledge, developing treatments and improving public health.

Key Concepts in Health Research

- The Scientific Method used a systematic approach to answering research questions through hypothesis testing and experimentation.

- Informed consent ensures participants understand the study, its purpose, risks and benefits before enrolling.

- Control groups are used to compare results with those of a test group, helping identify treatment effects.

- Bias and confounding are factors that can skew results, which researchers manage to ensure reliable findings.

- Blinding reduces bias by keeping participants and/or researchers unaware of who receives the treatment.

- Efficacy vs. Effectiveness: Efficacy is the effect in controlled trials, while effectiveness reflects real-world results.

Chapter Two

Ethical Principles in Health Research

E thics is about the moral rules that guide how people and organisations behave. It helps us decide what is right or wrong, based on values like fairness, responsibility and respect for others. In re-search, ethics is about protecting the rights, safety and well-being of people who take part in studies[9].

Research ethics guidelines have been created over time, often because of past cases where researchers treated participants unfairly or harmed them. Sadly, there have been many examples of unethical re-search, which has caused the public to lose trust in science. One well-known example happened during World War II in Germany when Nazi 'scientists' conducted experiments on prisoners. After the war and during the Nuremberg trials these terrible events came to light. Subsequently, the first standard of health research was created.

The Nuremberg Code[10] set basic rules like people must choose to take part voluntarily and give informed consent. The document, along with another set of guidelines called the Declaration of Helsinki[11] (created by the World Medical Association), established ethical principles for research involving people.

The Syphilis Study at Tuskegee: A Dark Chapter in Medical Research

The Syphilis Study at Tuskegee, conducted by the U.S. Public Health Service (USPHS) between 1932 and 1972, is one of the most infamous examples of unethical medical research in history[12].

The study sought to observe the natural progression of untreated syphilis in African American men in Macon County, Alabama. Of the 600 participants, 399 had syphilis, while 201 served as a control group. The men were neither informed of their diagnosis nor offered treatment, even after penicillin became the standard cure for syphilis in 1947.

The study's name reflects its geographic location near Tuskegee, Alabama, and its connection to Tuskegee Institute (now Tuskegee University). While the university provided facilities and employed a nurse who acted as a liaison, it played no part in designing the study or in the unethical decisions made. The harm and deception were entirely the responsibility of the USPHS.

Fortunately, this led to major changes in how research was undertaken. For example, new rules were put in place to protect participants and stop such unethical behaviour from happening again. One important result was the creation of the Belmont Report in 1979[13], which laid out key ethical principles for research with human participants, like respect, doing good and fairness.

This study caused immense harm and led to a loss of trust in the medical system, particularly among African American communities. In 1997, President Bill Clinton issued an apology on behalf of the U.S. government to the survivors and their families. Today, Tuskegee University works to honour the victims by promoting ethical research through its National Centre for Bioethics in Research and

Health Care, ensuring that lessons are learned from this tragic chapter in history.

Now, informed consent is a key part of research ethics. It makes sure that people taking part in a study understand what the study is about, any risks involved and that they can leave the study at any time. This helps protect people from harm and respects their right to make their own decisions, building trust between researchers and participants.

Research ethics provide a necessary set of rules to ensure studies are done in a way that respects people's dignity and maintains the honesty of the research. By following these guidelines, researchers make sure their work benefits society while protecting the rights and well-being of all participants.

Fundamental Ethical Principles: Respect for Persons, Beneficence and Justice

Three main ethical principles guide research: respect, doing good (beneficence) and fairness (justice)[13]. These principles show how seriously the research community takes its duty to protect participants and make sure studies are done fairly and openly.

Respect is the most important principle in ethical health research. It means treating all participants as independent individuals who have the right to decide if they want to join a study. To help them make an informed decision, they need to be given clear and detailed information about the study, like its purpose, what they'll need to do, any possible risks or benefits and the option to leave the study at any time.

Informed consent is important as it ensures that no one is pressured or misled into participating. Researchers must honour a person's choice to say no. Respecting people's autonomy protects their rights and ensures they can make the best choice for themselves.

Beneficence means that researchers should always act in the best interest of participants, putting people above developments in science. Before starting a study, researchers must carefully think about the possible benefits and risks.

For example, in a clinical trial, the benefits of a new treatment might include a reduction in symptoms. But there are also risks: there could be side effects and there is a chance the treatment won't work. Researchers must make sure the risks are minimised as much as possible and that the benefits are worth it. This requires careful planning, regular checks and quick action if anything unexpected happens.

Justice is the third principle, and it focuses on fairness in health research. It means that all participants should be treated equally, no matter their background, ethnicity, sex/gender, sexual orientation or personal situation.

No group should face more of the risks and no group should be left out of the potential benefits. For example, if a treatment could help many people, the study should include a diverse group of participants. This is important because treatments can sometimes work differently for different people and leaving some groups out could lead to unequal access to new treatments. Justice also means that the burdens of research, like time, side effects or risks, shouldn't fall more on one group than another.

Dr Paul Science Says
'Balancing benefits and risks are essential, not only to protect participants but also to advance science.'

Informed Consent

The informed consent process should be seen as a conversation between the participant and researcher. Participants should feel free to ask questions about the study. They can ask for more information or raise concerns at any time. Researchers must explain the study clearly, using simple language without technical terms.

For example, if a clinical trial is testing a new drug, people need to know how the drug will be given, how long the trial will last and what side effects might happen. This openness is important because it helps people weigh the potential benefits against the risks and decide if they want to take part. It's important that participants aren't rushed into deciding. They should have plenty of time to think it over and are encouraged to talk about the study with family, friends or doctors if they want. However, in real life, this ideal process isn't always followed perfectly due to practical challenges.

For example, in stroke research, where I used to work, patients sometimes needed to be enrolled in studies within hours of their stroke. This meant the information had to be presented very quickly and participants had less time to consider their decision. These situations bring additional challenges, so informed consent can be tailored to the type of research being done. In cases like emergency care studies, a more immediate decision is required.

Example 1: Sleep Deprivation Study

A psychology researcher wants to investigate the effects of sleep deprivation on cognitive performance. For example, how a lack of sleep might impact thinking and memory. The study involves university students as participants and, like any ethical research, it must follow a careful informed consent process to ensure that the students fully understand what they are agreeing to and that their participation is entirely voluntary.

The first step in this process is for the researcher to explain the study's purpose to the students. In this case, the goal is to understand how sleep deprivation might affect cognitive functions such as memory recall, concentration and problem-solving abilities. The researcher also outlines the procedures that the students will follow. For example, they will be asked to stay awake for a certain number of hours and complete a series of cognitive tests at various points throughout the study.

These tests might assess their ability to remember lists of words, solve puzzles or react to different stimuli, all of which help to measure how well their brains are functioning under conditions of sleep deprivation. It's important that the students not only understand what the study involves but also that they are aware of any potential risks. The researcher encourages them to ask questions about the risks, such as the possibility of feeling fatigued, drowsy or irritable because of staying awake for extended periods. It is essential for the researcher to be transparent about how these risks will be managed.

For instance, the students might be told that they will have access to a rest area after the study or that they can leave the study if the fatigue becomes too overwhelming. Open communication about these risks allows the students to weigh the pros and cons, based on their personal circumstances and tolerance for discomfort. Some students might need to think about how a lack of sleep could affect their academic work or social commitments, while others might want to discuss it with friends or family.

Throughout the informed consent process, the researcher makes it clear that participation is entirely voluntary. The students are reassured that they are under no obligation to take part and that, if they do decide to join the study, they can withdraw at any time without any negative consequences. Importantly, the researcher emphasises that withdrawing from the study will not affect the students' academic standing or their relationship with the university.

Once the students have had time to consider their options and decide to take part, they are asked to sign a consent form. Signing the form is a formal acknowledgment that they have made a voluntary, informed decision. However, the students can still ask questions or decide to withdraw from the study at any point if they feel uncomfortable or change their minds.

Example 2: Clinical Trial of a New Medicine

In a clinical trial of a new medicine, informed consent is both an ethical obligation and a legal requirement. The first step in this process involves providing potential participants with a detailed information sheet. This document contains clear and accessible information about the study's purpose. It will also specify how long the study will last and describe the procedures that participants will undergo, such as how the medication will be administered and how their health will be monitored.

The information sheet also outlines the potential benefits and risks. This is particularly important in clinical trials for new medicines, where the side effects of the drug may not yet be fully understood. In addition to detailing the new drug being tested, the information sheet will also mention any alternative treatments available for the condition being studied, so that participants can weigh up all their options before deciding whether to take part.

Once potential participants have received this information, they are encouraged to ask questions. The research team plays a key role in this part of the process, answering any concerns or uncertainties the participants may have about the study. It's vital that participants feel comfortable discussing the trial openly so they can gain a full understanding of what their participation will involve.

Questions might cover areas such as how their privacy will be protected, what happens if they experience side effects or what support is available if they decide to withdraw. Participants are then given time to review the information at their own pace. Once a participant has had time to review the information and is satisfied with their decision, they sign the informed consent form and can proceed.

Dr Paul Science Says
'Ethical research is inclusive research. True fairness means giving everyone a seat at the table.'

Privacy, Confidentiality and Data Protection

Protecting the privacy of participants and ensuring the confidentiality of their personal information is essential for both legal and ethical reasons. Participants place a great deal of trust in researchers when they agree to share sensitive information and it is the responsibility of the research team to safeguard that trust. This includes how information is handled, stored, shared and disposed of after the study is completed.

Researchers should always limit the amount of personal information they collect to only what is necessary. This reduces the potential risk of misuse but also respects participants' privacy by avoiding the collection of unnecessary data. The principle of data minimisation helps ensure that personal information is not over-collected and participants feel confident that their privacy is being respected.

To further protect participants' data, researchers must take active steps to secure the information they collect. This could involve using encryption for electronic data, which scrambles the information so that it can only be accessed by those who have the correct key or password. Encryption is a widely used method in health research because it helps prevent unauthorised access, especially if data is being transferred electronically or stored in online systems. Similarly, using strong passwords and regularly updating security protocols are also essential measures.

Researchers also need to carefully consider who has access to the data. Access should be restricted to authorised members of the research team who are directly involved in the study. These individuals should sign confidentiality agreements, which legally bind them to keep the data private and not share it with anyone outside the study. By limiting access in this way, researchers can significantly reduce the risk of data breaches and ensure that participants' personal information remains confidential. After the study is complete, researchers must have a clear plan for what happens to the data they have collected. In some cases, the data may be destroyed, particularly if it contains highly sensitive information that is no longer needed.

In other cases, the data may be anonymised, meaning all identifying information is removed so that the data can no longer be linked to individual participants. For example, researchers might remove participants' names, addresses and other personally identifiable information and replace them with unique codes or pseudonyms. This process of de-identification ensures that even if the data is shared with other scientists or used in further research, it cannot be traced back to the individuals who provided it.

In some studies, sharing data with other scientists may be necessary to validate findings or contribute to wider scientific knowledge. When this happens, researchers must still take care to protect participants' identities.

Data should be anonymised and de-identified before it is shared and strong encryption measures should be in place to prevent unauthorised access. Sharing data in this way allows for collaboration and further research while ensuring that the participants remain unidentifiable. This commitment to privacy and confidentiality is a fundamental aspect of ethical health research, helping to protect the rights and well-being of all participants.

Key Takeaways

- Research ethics safeguards the rights, safety and well-being of participants. These principles emerged as a response to historical instances of unethical research, leading to guidelines like the Nuremberg Code and Declaration of Helsinki.

- The core principles of ethical research include respect for persons, beneficence and justice.

- Informed consent is an interactive process where participants receive understandable information about the study, including benefits and risks. Participants should have the freedom to ask questions, take time to decide and withdraw at any point without repercussions.

Chapter Three

Research Question and Hypothesis

E very research study begins with a question that guides the investigation and shapes the entire project. This research question is an important starting point because it defines what the researchers want to discover or understand. The question must be clear, focused and directly relevant to the field of study. It should also address something that has not yet been fully answered by previous research.

While this might seem straightforward, formulating a good research question can be a complex process and is often one of the most challenging steps in designing a study[14]. Researchers typically begin by identifying a general area of interest or a specific problem they want to investigate. For example, they might be interested in exploring the effects of a new drug, understanding social behaviours or examining long-term health outcomes. Once they have a broad topic in mind, the next task is to narrow it down to a specific question that can be realistically answered within the scope of a single study. A research question needs to be specific enough to be answerable and broad enough to contribute to the area of study.

Dr Paul Science Says
'Never underestimate the power of a precise question. It's often the difference between notable research and a study that misses the mark.'

To illustrate this, let's take the example of social media use among teenagers. A researcher may be interested in how social media affects young people's mental health. However, to design a practical study, this needs to be refined into a focused question, such as: 'How do social media habits influence anxiety levels among teenagers during their high school years?' The question is specific, as it defines the population of interest (teenagers in high school), the behaviour being studied (social media habits) and the outcome being measured (anxiety levels). By narrowing the focus, the researcher has a question that is both relevant and researchable.

Another example could involve the testing of a new medication. Perhaps researchers wish to improve recovery times for patients undergoing knee replacement surgery. In this example they might start with a broad goal of assessing treatment options. However, a well-crafted research question would look more like: 'Does the addition of the new drug X improve recovery time in patients undergoing knee replacement surgery compared to the current standard treatment?' This question is clear and focused. It specifies the intervention (the new drug), the population (patients having knee surgery) and the comparison (standard treatment).

In some cases, research questions focus on long-term outcomes, which often require a different approach. For instance, if researchers wanted to understand the long-term health effects of different diets, they might ask: 'Over a 20-year period, what is the incidence of heart disease among vegetarians compared to non-vegetarians of similar demographics?' This question involves a long-term follow-up of two specific groups (vegetarians and non-vegetarians) and focuses on a particular outcome (incidence of heart disease). While the scope of the question is larger in terms of time and population, it is still specific enough to guide the design of a rigorous study.

Dr Paul Science Says
'A strong research question is like a compass — without it, even the best-designed study will lose direction.'

Understanding Hypotheses in Research

Once the research question is defined, the next step is to create a hypothesis. A hypothesis could be described as an educated guess about what the answer to the research question might be. It is based on what researchers already know, either from previous studies or their own observations, but it still needs to be tested through research to determine whether it is correct.

A well-formulated hypothesis is an essential element of any research study. It is the basis for the study design and the data collection methods. For a hypothesis to be useful, it must meet certain criteria. First, it must be testable. This means that researchers need to be able to gather data and evidence to either prove or disprove the hypothesis. Without this, the hypothesis would be purely speculative and not grounded in evidence. Second, the hypothesis must be specific. It should clearly outline what the researcher expects to find, leaving no room for ambiguity. For example, instead of a vague statement like 'a new drug will help patients feel better,' a specific hypothesis might state, 'patients taking the new drug X will experience a 20% faster recovery time compared to those taking the standard treatment'. This level of detail makes it easier to design a study that can accurately measure whether the hypothesis holds true.

The hypothesis must also be plausible. It should be grounded in existing knowledge and observations rather than a random or unsubstantiated guess. This doesn't mean the hypothesis must align perfectly with what's already known — it can propose something new or unexpected — but it should be a reasonable prediction based on the available evidence. This is why researchers often conduct thorough background research before developing their hypothesis, ensuring that it builds upon previous findings and doesn't contradict established science without a solid rationale.

There are generally two types of hypotheses that researchers work with: the null hypothesis and the alternative hypothesis. The null hypothesis proposes that there is no significant effect or relationship between the variables being studied. In other words, it suggests that any observed differences or changes are due to chance rather than a direct result of the intervention or factor being tested. For example, in a study testing a new medication, the null hypothesis might be that the drug has no more effect on recovery time than a placebo.

Alternatively, the alternative hypothesis suggests that there is a significant effect or relationship between the variables. This is what researchers are often hoping to prove. In the case of the new medication, the alternative hypothesis would propose that the drug does have a significant impact on recovery time compared to the placebo. The purpose of the research is to collect evidence that either supports the alternative hypothesis or leads to the acceptance of the null hypothesis, depending on what the data reveals.

Dr Paul Science Says
'In research, the null hypothesis keeps us grounded. It reminds us that proving something to be significant isn't a given—it's earned through rigorous testing.'

Throughout the study, data is gathered and analysed to determine which hypothesis is more likely to be correct. If the evidence strongly contradicts the null hypothesis, researchers may conclude that the alternative hypothesis is more likely to be true. However, it's important to note that research doesn't usually 'prove' a hypothesis in the strictest sense. It simply provides evidence that either supports or refutes the hypothesis, with varying degrees of confidence.

Even when a hypothesis is supported, researchers remain open to further testing and validation to confirm their findings. Understanding hypotheses is an important part of the scientific process because it demonstrates how researchers move from questions to answers. Hypotheses guide the direction of a study, helping to focus the research on specific predictions that can be tested through careful observation and data collection. Whether the hypothesis is confirmed or refuted, the research contributes valuable insights to the field, enhancing our understanding of the subject matter and often raising new questions for further investigation.

Variables and Outcomes

Variables are the different factors that researchers study to understand how they interact and affect results. They are like the building blocks of a study, helping researchers explore cause-and-effect relationships. There are two main types of variables: independent variables (the ones researchers change) and dependent variables (the ones researchers measure to see the effect).

Independent variables are the factors that researchers control or change in a study. They adjust these variables to see how they affect outcomes. For example, in a clinical trial comparing two drugs, the independent variable would be the type of drug each participant takes. Researchers control which drug is given to different groups to see its impact. Dependent variables are the results that researchers measure. These outcomes are expected to change based on the independent variable. For example, in the same drug study, researchers might measure blood pressure or symptom relief as dependent variables, showing how the type of drug affects these outcomes.

Outcomes in research are the specific results that researchers aim to measure to determine whether there is a relationship between the independent and dependent variables. These outcomes need to be clearly defined and measurable to ensure that the study produces reliable and interpretable results. For example, in a study looking at the effectiveness of a new exercise programme for people with diabetes, researchers might measure outcomes such as changes in blood sugar levels, weight or overall fitness.

These outcomes help determine whether the exercise programme is successful in improving health for people with diabetes. In another example, consider a study investigating whether a new diet helps lower cholesterol levels. The independent variable would be the type of diet that participants are asked to follow. Some participants might be assigned to follow the new diet, while others might continue with their usual eating habits. The dependent variable, in this case, would be the participants' cholesterol levels. Researchers would measure these levels before and after the study to see if the new diet leads to significant changes compared to the control group.

If outcomes aren't clearly defined or measurable, the study might not produce meaningful results. For example, in the study on diet and cholesterol, cholesterol levels could be measured through blood tests. This provides an objective way to see if the diet had any effect. If researchers only asked participants how they felt without checking cholesterol levels, the study's conclusions would be much less reliable.

Understanding research questions, hypotheses, variables and outcomes helps to grasp the basics of how studies are planned and carried out. Each part builds on the other: a research question leads to a hypothesis, which then identifies the independent and dependent variables. These variables define the outcomes researchers will measure to see if their hypothesis is supported.

Key Takeaways

- The research question is the foundation of any study, defining its purpose and scope. It should be clear, specific and relevant to the field, addressing gaps in existing research.

- After defining a research question, a hypothesis is formulated as an educated guess that the study will test. A good hypothesis is testable, specific, and grounded in existing knowledge, guiding the study's design and data collection.

- Independent Variables are those manipulated by researchers to observe effects.

- Dependent Variables are the measured outcomes that show the impact of the independent variable.

- Clearly defined and measurable outcomes are crucial for reliable and interpretable study results.

Chapter Four

Study Design

Study Design 1: Observational Studies

O bservational studies are where researchers observe and analyse data without directly intervening or controlling the variables being studied. In other words, they study things as they naturally occur, without trying to change anything. This is different from interventional studies, where researchers actively introduce something to see how it affects the participants.

Observational studies are useful for understanding patterns, trends and relationships between different factors and health outcomes. They can help generate new ideas for future research and provide valuable information about real-world situations. However, because researchers don't have control over the variables, observational studies can't prove that one thing directly causes another — they can only show associations or correlations. There are three main types of observational studies: cohort studies, case-control studies and cross-sectional studies[15].

Cohort Studies

In a cohort study, researchers follow a group of people (called a cohort) over time to see how certain factors affect their health outcomes. The cohort is typically defined by a shared characteristic, such as age, occupation or exposure to a particular risk factor. Researchers collect data on the participants at the beginning of the study and then periodically throughout the study period, which can last for months, years or even decades[15].

Cohort studies can be either prospective (looking forward in time) or retrospective (looking back in time). Prospective cohort studies start in the present and follow participants into the future, while retrospective cohort studies use existing data to look back at what happened to participants in the past. An example of a cohort study would be following a group of smokers and a group of non-smokers over 20 years to compare their rates of lung cancer and other health problems.

Case-Control Studies

In a case-control study, researchers compare two groups of people: those who have a specific outcome or condition (cases) and those who don't (controls). They look back in time to identify factors that may have contributed to the development of the condition in the cases[15]. Researchers typically start by identifying a group of people with the condition of interest, such as a particular disease. Then, they select a control group of people without the condition who are otherwise like the cases in terms of age, gender and other relevant characteristics. They collect data on past exposures, behaviours and other factors that may be related to the condition.

Case-control studies are useful for studying rare conditions or outcomes that take a long time to develop. However, they can be prone to biases. For example, recall bias — where participants may not accurately remember past exposures — and selection bias — where the cases and controls may not be truly comparable. An example of a case-control study would be comparing the past diet and lifestyle habits of women with and without breast cancer to identify potential risk factors.

Cross-Sectional Studies

In a cross-sectional study, researchers collect data on a group of people at a single point in time. They measure various factors and outcomes simultaneously, providing a snapshot of the population at that moment. Cross-sectional studies can be used to assess the prevalence of a condition, describe characteristics of a population or explore relationships between different variables[15].

However, because cross-sectional studies only look at one point in time, they can't establish the temporal relationship between cause and effect. In other words, they can't tell us which came first — the exposure or the outcome. An example of a cross-sectional study would be surveying a sample of adults to determine the prevalence of diabetes and its relationship to factors like obesity, diet and physical activity.

> *Dr Paul Science Says*
> 'Each study design has its strengths and limita-
> tions. The key is knowing when and how to use
> each type to get reliable, relevant insights.'

Study Design II: Interventional Studies

Interventional studies, also known as experimental
studies, are a type of research where researchers actively introduce a treatment, intervention or change
to see how it affects the participants. This is different
from observational studies, where researchers simply observe and analyse data without directly ma-
nipulating any variables. In interventional studies,
researchers typically compare two or more groups of
participants: those who receive the intervention (the
treatment group) and those who don't (the control
group)[15].

These types of study allow researchers to understand
the cause-and-effect relationship between the intervention and the outcomes. Interventional studies are
essential for testing the safety and effectiveness of
new drugs, medical devices, procedures and other
healthcare interventions. They provide the highest

level of evidence for guiding clinical practice and decision-making. Clinical trials are an example of interventional studies.

Dr Paul Science Says
'The controlled environment of interventional studies makes them a gold standard for testing new treatments. However, proper design here is essential for meaningful results.'

Key Takeaways

- In observational studies, researchers observe and analyse data without altering variables.

- In cohort studies, researchers follow a group (cohort) over time to study health impacts related to specific characteristics or exposures.

- In case-control studies, researchers compare individuals with a specific outcome (cases) to those without it (controls) to identify contributing factors.

- In cross-sectional studies, researchers measure factors and outcomes at a single point in time as a snapshot.

- Interventional studies are when researchers actively introduce a treatment or intervention to study its effects. They are designed to identify cause-and-effect, providing high-quality evidence for clinical decisions.

Chapter Five

Types of Health Research

Health research is a vast field and includes many different types of studies. Each type is designed to improve our understanding of health and to enhance medical treatments, public health strategies and healthcare systems. It's helpful to think of health research as being under a big umbrella, with each type of research addressing a different aspect of human health. Each plays a unique role in advancing our knowledge and finding new ways to prevent or treat disease.

One of the foundational types of health research is basic research, which you could think of as the detective work of science. In these studies, researchers focus on the smallest parts of our bodies, like cells and genes, to understand how they work.

By studying these building blocks of life, scientists can identify what goes wrong in various diseases. For example, by understanding how certain cells malfunction in cancer, researchers can look for ways to correct or stop this process. Basic research may not immediately produce a new treatment, but it's the groundwork for everything that follows[16].

Translational Research

Translational research is often seen as the bridge linking scientific discoveries with real-world patient care. Sometimes referred to as 'bench-to-bedside' research, it takes findings made in the lab and turns them into new treatments or medical practices that can be used with patients.

For example, scientists might discover how a specific gene functions, which could then inspire the development of a new drug. This drug would go through clinical trials before eventually becoming a treatment option for patients. Translational research ensures that scientific breakthroughs don't stay within the lab but are actively put to use to improve patient care.

Dr Paul Science Says
'Think of translational research as a bridge – it's the essential link that turns lab breakthroughs into real-world treatments.'

Clinical Research

This type of research includes clinical trials, where new drugs, medical devices or treatments are tested to see if they are both safe and effective. In a clinical trial, researchers might test a new medication for high blood pressure, observing how it affects patients compared to those receiving a standard treatment or a placebo. The results of these trials determine whether the new treatment can be approved for wider use. Clinical research is how theoretical discoveries in the lab become actual treatments available to the public.

Working on a Phase I Clinical Research Unit

In my early days as a research technician in an early-phase clinical research unit, my work was organised into 'rounds.' At 09:00, the first participant (01) would receive their medication (usually a novel study drug).

The doctor and nurse administering the medication would then move to Participant 02 and give their dose at 09:05, and so on. Each round involved moving from participant to participant, performing observations and procedures as outlined in the study protocol. Every step had to be precisely timed. In most cases, they were spaced just five minutes apart.

If we had any issues (such as difficulties collecting blood), it could delay the next participant. In those moments, teamwork was essential. There were often other colleagues on hand, stepping in to keep things running smoothly. It was a busy, structured environment. I learned a great deal working in this way and it set me up for my career!

Beyond these broad categories, there are more specialised forms of research that focus on specific aspects of health and human behaviour.

Behavioural Studies

Behavioural studies investigate how people's behaviours affect their health. For example, they might look at how stress influences eating habits or how physical activity impacts mental health. These studies help us understand how lifestyle choices can affect long-term health outcomes. They are help inform public health campaigns aimed at encouraging healthier behaviours.

Genetic Studies

Genetic studies focus on the role that genes play in health and disease. By studying the genetic makeup of individuals, researchers can identify the genetic factors that increase the risk of certain conditions, such as heart disease or cancer. This knowledge can lead to personalised treatments that are tailored to a person's specific genetic profile.

Observational Studies

Observational studies take a different approach by monitoring participants without interfering with their lives. Researchers track a group of people over time to see how different health outcomes develop based on their behaviour or environment. For instance, an observational study might follow people who smoke and those who don't, looking at how rates of lung disease differ between the two groups. This type of research helps identify risk factors for diseases and gives insight into how different lifestyles or exposures impact long-term health.

Dr Paul Science Says
'In science, no question is too small. Observational studies show us that sometimes, just watching and learning can reveal life-changing insights.'

Physiological Studies

Physiological studies delve into the workings of the body, exploring how different systems — such as the cardiovascular or respiratory systems — function and interact. These studies are essential for understanding how the body responds to different stimuli, whether it's physical activity, stress or illness. For example, a physiological study might look at how the heart and blood vessels respond during exercise, providing insights into how to improve cardiovascular health.

Prevention Studies

Prevention studies aim to stop diseases before they start. They test methods such as vaccines, lifestyle changes or early medical interventions to see how well they prevent the onset of specific conditions. For example, researchers might study whether a new vaccine can prevent the spread of a virus or whether a specific diet can lower the risk of developing diabetes. It should be noted that in the case of a vaccine study, this would also fall under the category of a clinical trial, so sometimes studies fall across multiple categories. Again, this would need to be clearly defined in the study protocol.

> *Dr Paul Science Says*
> 'Prevention research reminds us of a simple truth: the best cure is to prevent the illness from occurring in the first place.'

Epidemiological Research

Moving from individuals to larger groups, epidemiological research looks at health trends across populations. Epidemiologists study how diseases spread, what factors contribute to their development and who is most at risk. This type of research is vital for public health planning, as it helps identify the causes of diseases and enables health authorities to prevent or control outbreaks. For example, during the COVID-19 pandemic, epidemiologists were involved in tracking how the virus spread and in identifying factors that increased the risk of severe illness, such as age or pre-existing conditions.

Public Health Research

Finally, public health research seeks to improve health at the population level, often by combining elements of all the other types of research. It might involve studying the impact of health policies, evaluating large-scale health interventions or exploring how social factors like income or education influence health outcomes. Public health research plays a key role in shaping government policies and health recommendations that benefit entire communities.

Health Services' Research

These types of research study focus on the healthcare system itself, investigating how healthcare is delivered and exploring ways to improve its efficiency, accessibility and effectiveness. It examines issues like waiting times in hospitals, the quality of patient care and the best ways to allocate limited healthcare resources. The goal is to ensure that the healthcare system meets the needs of the population as effectively as possible.

Clinical Trials – A Closer Look

Clinical trials, particularly randomised controlled trials (RCTs), are often described as the gold standard of health research. They play a vital role in determining whether new treatments, drugs or medical devices are safe and effective for use in patients. The reason they are the gold standard is because these trials are carefully designed to ensure that the results are as reliable and unbiased as possible.

One key aspect of this design is randomisation, where participants are randomly assigned to receive either the new treatment or a placebo (such as a sugar pill with no active ingredients).

This random assignment helps to eliminate any potential bias that could come from differences between participants, such as age, gender or underlying health conditions. By comparing the outcomes between the group receiving the treatment and the group receiving the placebo, researchers can determine whether the new treatment is genuinely effective or if the results are due to chance.

Clinical trials are conducted in phases, with each phase designed to answer different questions about the treatment being studied. These phases ensure that the treatment is carefully assessed for safety and effectiveness before it becomes widely available.

In phase I, the drug or treatment is tested in humans for the first time. A small group of healthy volunteers or patients (typically between 20 and 80 people) receive the treatment. The primary goal in this phase is to evaluate the safety of the drug, determine the correct dosage and identify any potential side effects. Phase I trials are usually not randomised because the primary focus is on understanding how the drug behaves in the human body. Researchers need precise control over who receives the treatment and in what dose to ensure that they can accurately assess its safety and potential side effects.

Once the treatment has passed this initial safety assessment, it moves into phase II. This phase involves a larger group of participants, usually between 100 and 300 people, who have the disease or condition the treatment is intended to address. In phase II, researchers focus on determining how effective the treatment is and continue to monitor its safety. This is where they begin to explore whether the treatment is having the desired effect on the disease or condition. They also gather more detailed information about side effects. Randomisation is typically introduced in this phase, as researchers compare the new treatment with a placebo or an existing standard treatment to see if it offers any significant advantages.

In phase III, the trial expands to a much larger group of participants, often in the many thousands. Some phase III trials have had over 40,000 participants involved. The goal here is to compare the new treatment to the current standard of care or to a placebo on a large scale, providing a clearer picture of its effectiveness and any additional benefits it may offer.

Researchers continue to monitor side effects and the data collected from this phase is critical for regulatory approval. In the UK, for example, approval might come from the Medicines and Healthcare products Regulatory Agency (MHRA), while in the US, the Food and Drug Administration (FDA) is responsible for this role. The European Medicines Agency (EMA) oversees drug approvals across the EU.

Phase III trials are often considered the most decisive in determining whether a treatment is safe and effective enough for widespread use. After a treatment has been approved by regulators, the medicine can be licensed and may be used within the population.

Ongoing safety is monitored in phase IV trials. These trials involve thousands of participants and focus on monitoring the long-term effects of the treatment in the general population. Phase IV trials help to ensure that any rare or long-term side effects that didn't appear in earlier phases are identified. They also provide further data on the treatment's efficacy over extended periods, ensuring that it remains safe and beneficial for patients over time.

Dr Paul Science Says
'To advance medicine, we need both innovation and caution – clinical trials balance the excitement of new discoveries with a commitment to safety.'

Key Takeaways

- Health research has many types, each targeting different aspects of health and disease.

- Basic research conducted at the cellular or genetic level, builds our understanding of the body's smallest components.

- Translational or 'bench-to-bedside' research transforms lab discoveries into practical treatments.

- Clinical research, particularly through clinical trials, rigorously tests new treatments and interventions for safety and effectiveness before they are widely adopted.

- Health services research focuses on healthcare delivery, aiming to improve efficiency, accessibility and patient care within healthcare systems.

- Clinical trials, especially randomised controlled trials (RCTs), offer the most reliable results by rigorously testing new treatments through phased assessments (I–IV), ensuring safety and effectiveness at each stage.

Chapter Six

Population and Sampling

Defining the Target Population

T he target population is the group of people that researchers want to learn about in their study. This group could be large, like all adults over 18. Or it could more specific, like women diagnosed with breast cancer or children with asthma. Defining this group carefully helps make sure that study results are useful and relevant for the people the researchers are interested in.

When choosing a target population, researchers think about factors like age, as health conditions and responses to treatments can vary widely between ages. For example, a study on a new arthritis treatment might focus on older adults. This is because older adults are more likely to have arthritis. In contrast, a study on childhood asthma would naturally focus on children. By choosing a specific age range, researchers ensure that the study results will be relevant to the people most affected.

Sex is another important factor to consider in research. Some health issues, like prostate cancer or ovarian cancer, only affect men or women, so researchers focus on the relevant group. In other cases, researchers may want to see if a treatment works differently for men and women. For example, a study on heart disease might look at whether women and men respond differently to a medication, as heart disease can affect each group differently.

Health status is also important when choosing who to study. Some studies focus on people who already have a condition, like diabetes or high blood pressure, to see how a treatment works for them. In these cases, only people diagnosed with the condition are included. Other studies, however, may look to prevent disease and so they might include healthy people who are at risk of developing a condition. For example, a study on a new vaccine would focus on healthy people who could be at risk of infection.

Geographic location can also be important in choosing the target population. Researchers might focus on a specific region or country, especially if they are studying diseases or health issues that are more common in certain areas. For example, studies on malaria would likely focus on areas where malaria is common, like parts of Africa or Southeast Asia. A study on air pollution's impact on breathing might focus on people living in cities with high pollution levels. Choosing a location makes sure the study relates to the unique environment and lifestyle factors that could affect health.

Other factors, such as lifestyle, income, genetics, occupation and even access to healthcare, can also help define the group researchers want to study. For example, lifestyle habits are often central to studies exploring behaviours linked to health. A study on quitting smoking, for example, would include people who currently smoke or who have recently quit. This is because they are most likely to benefit from or need support with smoking cessation programs. Similarly, studies on the impact of physical activity on heart health might focus on people with varying levels of exercise habits, from those who are highly active to those with a more sedentary lifestyle.

Income and socioeconomic status are also important, as they can affect access to healthcare, nutrition and overall health. Research on the effects of economic status on health outcomes might look at individuals from different income groups.

Genetics is another key factor, especially for conditions that can run in families, like Alzheimer's disease or certain types of cancer. For example, research on genetic risks for Alzheimer's might include people with a family history of the disease, as they may be more likely to carry specific genetic markers that the study wants to explore. Or research looking into hereditary cancers might involve individuals with genetic mutations.

Occupation and work environment may also play a role. Studies on respiratory health might focus on people working in industries with higher exposure to pollutants, such as construction or mining.

Finally, access to healthcare services can influence health outcomes and may shape the target population for some studies. Research on managing chronic conditions might focus on individuals with limited access to regular healthcare, as their experiences may differ significantly from those with easy access to medical support.

> *Dr Paul Science Says*
> 'A well-defined target population is essential. Always ask: Who exactly are we studying, and why?'

Sampling Methods: Probability and Non-Probability Sampling

In health research, it is often impractical or impossible to study every individual within the target population. To overcome this, researchers select a smaller group of individuals, known as a sample, to participate in the study. The aim is to choose a sample that accurately reflects the larger population so that the results of the study can be generalised.

In other words, the findings from the sample should give researchers insights that apply to the broader group. To achieve this, researchers use different sampling methods, each with its advantages and challenges[17]. There are two main approaches to sampling: probability sampling and non-probability sampling.

Probability sampling is a method in which participants are randomly selected from the target population, ensuring that each individual has an equal chance of being chosen. This random selection helps to minimise bias and increases the likelihood that the sample will be representative of the larger population. In turn, this makes the study's results more reliable and applicable to the population[17].

One common technique within probability sampling is simple random sampling. In this method, every individual in the target population has an equal chance of being selected to participate. This could be achieved by assigning each person a number and then using a random number generator to select participants. Simple random sampling is a straightforward and effective way to create a representative sample, but it can be difficult to implement in large or geographically dispersed populations[17].

Stratified random sampling is another probability-based approach. In this method, the target population is divided into subgroups, or strata, based on key characteristics such as age, gender or ethnicity. Participants are then randomly selected from each subgroup. This approach ensures that each important subgroup is properly represented in the sample, which can lead to more accurate and meaningful findings. For example, in a study on heart disease, researchers might use stratified random sampling to ensure that both men and women are adequately represented, given that heart disease can affect the sexes differently.

Cluster sampling is another technique used when the population is large or spread out over a wide area. Instead of selecting individuals directly, researchers divide the population into clusters — often based on geographic location or other relevant factors — and then randomly select a few clusters. Everyone within the selected clusters is included in the study. This method is useful for large-scale studies because it reduces the logistical challenges of reaching participants, but it can introduce some bias if the clusters themselves are not truly representative of the entire population[17].

On the other hand, non-probability sampling does not involve random selection. Instead, participants are chosen based on convenience, availability or specific criteria relevant to the research question. While non-probability sampling is often easier and quicker to implement, it carries a higher risk of bias. This means that the sample may not fully represent the target population. However, in some cases, non-probability sampling is the only feasible option, especially when it is difficult to access a large, randomised pool of participants.

One of the most common forms of non-probability sampling is convenience sampling. As the name suggests, participants are selected based on their accessibility or willingness to take part in the study. For example, a researcher might recruit participants from a nearby hospital or university because it is convenient. While this method can save time and resources, it risks producing results that only reflect the experiences of a specific, possibly unrepresentative group.

Another non-probability technique is purposive sampling, where participants are selected based on specific characteristics that are important to the research. In this approach, researchers deliberately choose individuals who meet particular criteria relevant to the study. For example, in a study on rare diseases, researchers may need to use purposive sampling to find participants who have the condition being studied. While purposive sampling is useful for targeting specific groups, it may limit the ability to generalise the results to the broader population[17].

Snowball sampling is another non-probability method. It is often used in studies where the target population is difficult to reach or identify. In this technique, researchers start by recruiting a small group of participants who meet the study criteria.

These initial participants are then asked to refer others who might also qualify for the study. As more participants are referred, the sample grows, like a snowball gathering in size as it rolls downhill. While snowball sampling can be useful for accessing hard-to-reach groups, it can also introduce bias, as the sample depends heavily on the social networks of the initial participants[17].

The choice between probability and non-probability sampling largely depends on the study's goals, the resources available and the characteristics of the target population.

Probability sampling methods are generally preferred when the goal is to obtain a representative sample that allows for generalisable results. However, in some cases, non-probability sampling may be more practical, especially when working with specialised or hard-to-reach populations.

Dr Paul Science Says
'Sampling may sound simple, but choosing the right sample can make or break your study's credibility.'

Sample Size Determination

When planning a study, determining the number of participants required is an essential step, as this directly influences the reliability and validity of the results. This is known as calculating the sample size. Having the right number of participants ensures that if there is a real difference or relationship between the factors being studied and that it can be detected with confidence.

An appropriate sample size allows researchers to make accurate conclusions about the data, reducing the likelihood that the results are due to chance[18].

A larger group of participants generally leads to more precise results. This is because individual differences or variations are averaged out in larger samples. However, increasing the number of participants comes with practical challenges, such as higher costs, more time spent recruiting, managing participants and the need for additional resources to conduct the study. Therefore, researchers must find a balance between obtaining enough participants to ensure reliable findings and managing the practical constraints of time, budget and logistics[18].

Study design impacts the sample size that may be required. Studies with complex designs or those looking at multiple variables may require more participants to detect meaningful effects. In addition, the 'effect size' — the effect researchers expect to see — is also relevent. If researchers are studying something with a large and obvious effect — such as a treatment that drastically improves a health outcome — they may need fewer participants to observe that effect. On the other hand, if the expected effect is small or subtle, a larger sample is required to detect it reliably[18].

The level of certainty researchers want about their results also affects the sample size. This is called statistical significance. It indicates how confident researchers are that the results they observe are real and not due to random variation. To achieve statistical significance, researchers set a threshold expressed as something called a 'p-value'. A lower p-value (such as 0.05) means researchers want to be very sure about the results, which typically requires a larger sample size to provide that level of certainty[18].

Another factor is the amount of error researchers are willing to accept in their findings, known as the margin of error. No study can ever be completely free from error, but researchers aim to minimise it as much as possible. A smaller margin of error requires a larger sample size, as it reduces the likelihood that random variations will distort the results.

For example, if a study is investigating a new diabetes medication, researchers might aim for a very low margin of error to ensure that the drug's effects are accurately measured. This would necessitate recruiting more participants to reach that level of precision[18].

Ultimately, calculating the sample size is a balancing act. If a sample size is too small, the study might not have enough power to detect significant effects. Conversely, an unnecessarily large sample size could waste resources without adding significant value to the findings. To help determine the right sample size, researchers often conduct a 'power analysis'. This is a statistical method that estimates how many participants are needed to detect an effect of a certain size with a given level of confidence[18].

> *Dr Paul Science Says*
> 'Your sample doesn't have to be huge; it just needs to be representative of the people who matter for your study's goals.'

Inclusion and Exclusion Criteria

Eligibility criteria are important because they define who is eligible to participate in the study and who is not.

Inclusion criteria outline the specific characteristics that participants must have to be part of the study. These criteria might include factors such as age, gender, health status or even willingness to follow the study protocols. For example, in a study looking at the effects of a new asthma treatment, the inclusion criteria might specify that participants need to be between the ages of 18 and 65 and have been diagnosed with moderate to severe asthma.

These requirements help ensure that the study is focused on the group most relevant to the research question and they also increase the likelihood that the results will be meaningful.

Exclusion criteria, however, list the factors that would prevent someone from participating. These criteria are used to exclude individuals whose characteristics or conditions might interfere with the results or put them at risk. For example, if the asthma study includes people who are taking certain medications that could interact with the new treatment, those participants might be excluded to keep the participants safe.

Excluding individuals who may have certain vulnerabilities or health conditions helps reduce the likelihood of adverse effects and ensures that the treatment or intervention is being tested in a population that can safely participate. For example, a clinical trial for a new heart medication might exclude individuals with a history of severe allergic reactions to similar drugs to prevent potential harm.

Example Eligibility Criteria

Here are examples of inclusion and exclusion criteria for a pain control study. However, please remember that these criteria will always differ depending on the requirements of the study.

Inclusion Criteria

1. Participants must be between 18 and 65 years old.

2. Both male and female participants are eligible.

3. Participants must have a diagnosis of chronic pain for at least six months.

4. Participants must provide informed consent and be willing to comply with study procedures.

5. Participants must be able to read and understand English to ensure they can follow instructions and complete questionnaires.

6. Participants must have a BMI between 18.5 and 30.

7. Participants must not have used any investigational drugs in the past 30 days.

8. Participants must reside within a 50-mile radius of the study site to facilitate attendance at study visits.

9. Participants must have a specific type of chronic pain, such as osteoarthritis or fibromyalgia.

10. Participants must have been on a stable medication regimen for at least three months prior to the study.

Exclusion Criteria

1. Participants with a history of severe cardiovascular disease are excluded.

2. Participants taking anticoagulant medications, such as warfarin, are excluded.

3. Pregnant or nursing women are excluded to avoid potential risks to the foetus or infant.

4. Participants with a history of substance abuse within the past year are excluded.

5. Participants with severe psychiatric disorders, such as schizophrenia, are excluded.

6. Participants who have undergone major surgery within the past six months are excluded.

7. Participants with a known allergy to aspirin are excluded.

8. Participants who are immunocompromised are excluded.

9. Participants who smoke more than 10 cigarettes per day are excluded.

10. Participants currently enrolled in another clinical trial are excluded to avoid confounding results.

Key Takeaways

- Defining a target population is essential in research as it helps focus the study on the specific group most relevant to the research question, such as age, sex, health status or geographic location. This targeted approach improves the usefulness and relevance of the findings.

- Sampling is necessary when it's impractical to study an entire population.

- Calculating an appropriate sample size is

- important to ensure that results are statistically reliable.

- Inclusion and exclusion criteria differ by study type.

Chapter Seven

Data Collection & Management

Types of Data: Quantitative and Qualitative

In health research, there are two types of data: quantitative and qualitative. Each serves a different purpose in helping researchers understand health issues. Together, they provide a fuller picture of the topic being studied. Quantitative data are numerical and can be counted, measured or analysed statistically. They provide concrete information about quantities, amounts or frequencies.

This means this type of data is useful for identifying patterns, measuring changes and drawing comparisons. For example, quantitative data in health research might include a patient's blood pressure readings, cholesterol levels, the number of hospitalisations or responses to a survey scored on a scale from 1 to 5. These numbers allow researchers to assess the effectiveness of a treatment, track health trends over time or evaluate how widespread a condition might be in a population[19].

Qualitative data, however, are non-numerical and focus on capturing the qualities, characteristics and experiences of participants. These data offer insights into people's thoughts, feelings and behaviours. In health research, this might involve open-ended interview questions where patients describe their experiences with a particular treatment or written feedback on how a condition affects their daily lives.

Unlike quantitative data, which can be analysed through statistical methods, qualitative data are typically analysed by identifying themes, patterns and meanings that emerge from people's responses. While it is more subjective, qualitative data offer depth and context, helping researchers understand the human side of health issues that numbers alone cannot fully capture[19].

Both quantitative and qualitative data are valuable, but they serve different functions in research. Quantitative data provide the statistical foundation needed to measure trends, test hypotheses or make predictions about health outcomes. They help answer questions such as 'How many people benefit from this treatment?' or 'What is the average change in cholesterol levels after taking this medication?'

However, quantitative data alone may miss the more personal, emotional or experiential aspects of health. This is where qualitative data becomes important, as they help researchers explore 'how' and 'why' certain health outcomes occur. For instance, qualitative data might reveal why some patients prefer one treatment over another or how living with a chronic illness affects their quality of life[19].

> *Dr Paul Science Says*
> 'Remember, people aren't numbers. Qualitative data keeps the human element front and centre in health research.'

Some health studies focus exclusively on one type of data. For instance, a clinical trial might rely solely on quantitative data to measure the effectiveness of a new drug by collecting and analysing health indicators such as blood pressure or heart rate. In contrast, a study exploring patients' experiences with mental health services might focus entirely on qualitative data.

Nowadays, many studies use a mixed-methods approach, combining both types of data. This approach allows researchers to get the best of both worlds. For example, in a study on a new cancer treatment, researchers might first collect quantitative data to measure tumour reduction or survival rates. They could then follow up with qualitative interviews to understand how patients feel about the treatment, how it affects their everyday lives and whether they experience any side effects that might not be captured by the numbers alone.

Data Collection Tools and Methods

In health research, collecting accurate and relevant data is essential to answering the research question and achieving reliable results. To gather this data, researchers use a variety of tools and methods, carefully chosen based on the study's design, the type of data needed and the characteristics of the participants.

Each method offers distinct advantages depending on the nature of the study, whether it involves gathering numerical information or exploring people's thoughts and behaviours.

One of the most common tools for data collection is the questionnaire or survey. They can be delivered in different formats, including paper forms or online. The advantage of using questionnaires and surveys is that they can efficiently collect data from large groups of people in a relatively short amount of time.

Interviews are another widely used method for data collection. Unlike surveys, interviews involve a one-on-one conversation between the researcher and the participant. These can be structured, where the researcher follows a specific set of questions, or semi-structured, where there is a general guide but flexibility to explore certain responses in more depth.

Unstructured interviews allow the conversation to flow naturally, giving participants the freedom to share their thoughts in an open-ended manner. Interviews are particularly valuable when researchers want to explore personal experiences, emotions or detailed opinions, as they provide the opportunity to delve deeply into the participant's perspective.

Focus groups are a method that involves guided discussions with a small group of participants who share common characteristics or experiences relevant to the research. The interaction between group members often sparks deeper discussion and allows researchers to observe how participants think and behave in a social context. Focus groups are commonly used in qualitative research to explore themes such as attitudes, beliefs or reactions to specific topics. For instance, a focus group might be used to gather feedback on a new public health campaign, where participants discuss how they perceive the messaging and whether it resonates with their community.

Observation is another important method used in qualitative research, particularly when researchers are interested in behaviours, interactions or environments. In qualitative observational studies, researchers watch and record participants in their natural settings, either directly or through video recordings.

Physical measurements are often used in studies that require objective numerical data related to participants' health, such as clinical trials. This method involves measuring various physical characteristics such as height, weight, blood pressure or lung capacity.

Medical records provide another source of data. They contain a wealth of information on participants' health history, diagnoses, treatments and outcomes. Researchers can extract this data directly from hospital records, electronic health records or national health databases, such as hospital registries.

Using medical records allows researchers to study long-term health trends, track the effectiveness of treatments, or identify patterns in healthcare delivery. This method is particularly useful in large-scale studies where it may not be feasible to collect primary data directly from participants.

Dr Paul Science Says
'Data is only as good as the care we take in collecting it. Small errors can lead to big misinterpretations.'

Quiz: *Identify the Data Collection Methods*

Can you identify which type of data these methods will create?

- Questionnaires and Surveys

  - Quantitative

  - Qualitative

- Interviews

  a. Quantitative

  b. Qualitative

- Focus Groups

  a. Quantitative

  b. Qualitative

- Observations

  a. Quantitative

  b. Qualitative

- Physical Measurements

- a. Quantitative

- b. Qualitative

- Medical Records and Databases

  - a. Quantitative

  - b. Qualitative

Answers

1. a Primarily Quantitative, but can be Qualitative depending on the questions,

2. b Primarily Qualitative,

3. b Qualitative,

4. b Can be both, but often Qualitative,

5. a Quantitative,

6. a Quantitative.

Data Management and Quality Assurance

In health research, proper data management and quality assurance are essential for ensuring the integrity, accuracy and usability of the data collected during a study. These processes help maintain the reliability of the research findings and ensure that the data can be used confidently to draw valid conclusions. Data management and quality assurance work hand in hand to safeguard the study from errors, inconsistencies, and potential data loss[9].

Data Management

Data management refers to the organisation, storage, and protection of data throughout the study9. Secure and reliable data storage systems are vital in protecting the data from unauthorised access, theft or loss. Researchers often use encrypted databases or password-protected files to ensure that only authorised individuals can access the data. These measures are essential for maintaining confidentiality, particularly when dealing with sensitive health information. In addition, secure storage helps prevent accidental data breaches that could compromise the privacy of participants and the integrity of the research.

Backing up data regularly is another essential aspect of data management. Technical issues, system failures or even simple human errors can lead to data loss, which could be catastrophic for a research study. Regular backups ensure that a copy of the data is always available, safeguarding against potential losses. It is similar to making copies of important documents and storing them in a safe place — if something goes wrong, there is always a reliable backup to fall back on.

Ensuring accurate and consistent data entry is also key to maintaining the integrity of the research. When data is being inputted into a system, mistakes or inconsistencies can easily occur, especially in large studies. To minimise these risks, researchers often use standardised procedures and data entry forms. This ensures that all data is entered in a consistent format, reducing errors and making it easier to analyse later. Accurate data entry is the foundation of reliable research, as even small errors at this stage can lead to incorrect results or misleading conclusions.

Dr Paul Science Says

'In health research, data quality is non-negotiable. Reliable data means reliable conclusions, and that's what saves lives.'

Quality Assurance (QA)

Quality Assurance (QA) is the process by which researchers monitor and improve the quality of the data collected. This is vital for ensuring that the data is accurate, complete and useful for the research objectives. A key aspect of quality assurance is training the individuals responsible for collecting data[9].

Regular data audits are another important part of QA. Audits involve checking the data for any inconsistencies, missing values or implausible entries that might indicate errors. This process is akin to proofreading a document to catch mistakes before finalising it.

Conducting these audits regularly allows researchers to spot potential problems early, preventing small issues from becoming larger obstacles later in the study. Data cleaning is the next step in ensuring quality. After the data is collected, it is essential to go through it carefully to identify and correct any errors or discrepancies[9]. Finally, documenting any changes or corrections made to the data is important for maintaining transparency and traceability. Researchers often use version control systems or audit trails to record what changes were made, when they were made and why.

Key Takeaways

- Health research collects two primary types of data: quantitative (numerical, statistical data) and qualitative (non-numerical, experiential data).

- Both types offer unique insights — quantitative data for identifying patterns and measuring trends and qualitative data for understanding personal experiences and emotions.

- Combining quantitative and qualitative data in a mixed-methods approach offers a comprehensive view of health issues, validating statistical findings with personal experiences and vice versa.

- Data management involves planning, organising, storing and backing up data to maintain security, accuracy and accessibility.

- Quality Assurance ensures data reliability through training, audits and data cleaning.

Chapter Eight

Understanding Data in Research

W hen researchers complete their data collection, their next task is to organise and summarise the information in ways that show key patterns and trends. This is done using descriptive statistics: a set of methods that allow them to describe the basic features of the data.

These statistics help researchers understand what the data looks like, spot any obvious trends and identify anything unusual, such as outliers that don't fit with the rest of the data.

Descriptive Statistics

One key part of summarising data is looking at the 'middle' or 'typical' value, which tells us what's most common or average in the data. This is described as the measure of central tendency. The most well-known way to find this is by calculating the mean or average. To get the mean, you add up all the numbers in the group and then divide by how many numbers there are. This gives you a good idea of the overall 'centre' of the data. For example, if researchers are studying blood pressure in a group of patients, the mean would give an idea of the typical blood pressure in that group. Another measure is the median: the middle value when all the data points are arranged in order from smallest to largest.

The median is especially useful when there are extreme values in the data that might skew the mean. For example, if one patient has an unusually high or low blood pressure compared to the others, the median provides a clearer picture of the typical value in the group. Lastly, the mode is the value that appears most frequently in the data set. In some cases, this can be a helpful way to understand the most common outcome or characteristic in a sample.

Dr Paul Science Says
'Visuals like charts and frequency distributions make complex data easier to understand and share with others.'

In addition, researchers also look at measures of variability to understand how spread out their data is. The range, for instance, is the difference between the highest and lowest values in the data set. It provides a simple measure of the spread. Another important measure is the standard deviation, which tells researchers how much the data varies from the mean.

A small standard deviation indicates that most of the data points are close to the mean, while a large standard deviation means the data points are more spread out. For example, in a study on blood pressure, a large standard deviation might suggest that patients have widely varying blood pressure levels, whereas a small standard deviation would indicate that most patients have similar readings.

Researchers often use frequency distributions to visually represent their data. It can be presented in tables or charts, making it easier to see how the data is distributed. For example, a frequency distribution might show how many people fall into different age groups or how many patients have blood pressure readings within certain ranges. This helps researchers quickly grasp the overall shape of the data and spot any patterns or anomalies.

Inferential Statistics

While descriptive statistics help us see a summary of the data we have, inferential statistics let us make educated guesses about a larger group based on just the sample we studied. This type of analysis is important in health research because it helps scientists decide if what they found in a small group might also be true for many more people, especially when studying the effects of a treatment or intervention. One of the key tools in inferential statistics is hypothesis testing, which involves testing assumptions about the data. For example, researchers might test whether a new drug works better than the current standard treatment.

By using statistical tests, they can determine whether any differences observed in the study are likely to be real or simply due to chance. Another important aspect of inferential statistics is the use of confidence intervals. A confidence interval provides a range of values where the true value for the entire population is likely to fall. For example, if a study finds that a new drug lowers blood pressure by an average of 10 points, the confidence interval might suggest that the true reduction in the larger population is likely to be between 8 and 12 points. This range shows how precise the findings are and how confident they can be in the results.

Regression analysis is another tool that is used to examine relationships between variables. For example, in a study looking at the effect of weight loss on blood sugar levels, regression analysis could help researchers understand how much of an impact weight loss has on improving blood sugar control. This type of analysis can also help predict how changes in one variable might affect another, providing valuable insights into cause-and-effect relationships in health research.

Interpreting the results of a study requires careful consideration of several factors. Researchers need to look at the effect size, which refers to how large the observed effect is and the direction of the effect — whether it shows an improvement, decline or no change. For example, a study might show that a new medication reduces symptoms significantly or it might show only a slight improvement. Understanding the size of the effect helps researchers determine the practical importance of their findings. In addition, researchers assess the certainty of their findings by looking at statistical significance and confidence intervals.

Statistical significance indicates how likely it is that the results are real and not due to chance. Researchers often use the p-value (commonly less than 0.05) to decide this. If the p-value is below this threshold, they conclude that the observed effect is statistically significant and reflects a real difference rather than random variation. However, statistical significance alone doesn't tell the whole story, so researchers also compare their findings with previous studies to see if their results align with existing evidence. This helps validate the findings and provides a broader context for understanding their relevance.

Dr Paul Science Says
'Data doesn't speak for itself; it needs careful analysis to reveal its true meaning.'

Key Takeaways

- Researchers begin by organising and summarising collected data through descriptive statistics to identify key trends and patterns.

- Measures of central tendency, like the mean, median and mode, help determine the typical values within the data, while measures of variability, such as range and standard deviation, show how spread out the data points are.

- Beyond describing the data, inferential statistics allow researchers to make predictions about a larger population based on a sample.

- Hypothesis testing and confidence intervals are key components, helping determine whether results are meaningful and likely to apply more broadly.

- Statistical tests, like the p-value, indicate the likelihood that results are genuine rather than random. A statistically significant result (p-value < 0.05) suggests that the findings are meaningful, though they must be interpreted in the context of existing research.

Chapter Nine

Communicating Research Results

C ommunicating research findings is one of the final steps in the research process. It ensures that the knowledge gained from a study reaches those who can benefit from it, such as healthcare providers, policymakers, patients and the public. The way research findings are shared depends on the audience and the intended purpose, with various methods used to disseminate the information.

Dr Paul Science Says
'No research discovery is truly complete until it's shared with the world. How we communicate results determines the real-world impact of our work.'

Peer-Reviewed Publications

One of the most established ways to communicate research findings is through peer-reviewed publications. These are articles published in scientific journals that have undergone a rigorous review by independent experts in the field. Peer review is a key aspect of scientific publishing because it ensures that the research is thoroughly evaluated for its quality, validity and significance. The process is lengthy and competitive and not all submissions are accepted.

Researchers must first write a detailed manuscript following the journal's specific guidelines. This manuscript outlines the study's methods, results and conclusions. After submission, the journal's editor conducts an initial review to assess whether the paper is suitable for further consideration. If it passes this stage, it is sent to two or more experts in the relevant field for peer review.

During peer review, these experts carefully examine the research to evaluate how well the study was designed, the accuracy of the statistical analyses and the interpretation of the results. They will also assess how the research contributes to the field. Based on their assessment, the reviewers recommend what happens next. Either the manuscript is accepted or the reviewers' request that revisions are made to improve the manuscript's clarity and/or accuracy. In some cases, the journal may reject the manuscript altogether. If revisions are requested, the researchers must address these comments and resubmit the manuscript.

Once accepted, the manuscript undergoes final editing and formatting before being published in the journal. Once published, it becomes part of the scientific record, where it can be accessed by other researchers, healthcare professionals, patients and the public.

In addition to peer-reviewed journals, researchers frequently present their findings at conferences. They may present their findings through oral talks or as poster presentations; where visual summaries of the research are displayed for attendees to view and discuss. Conferences are a valuable platform for networking, fostering collaborations and staying up to date with the latest developments in a field.

Press and Media

Research findings are also communicated through press releases and media coverage. Universities or research institutions often collaborate with public relations personnal to create press releases that summarise a study's key findings in clear, accessible language. Press releases are picked up by journalists, who then write articles about the research. This type of communication helps raise awareness about important research and its potential impact on health, policy or society at large.

Social media and blogs have become increasingly popular tools for researchers to communicate their work. Platforms such as Threads, X (previously Twitter) and LinkedIn allow researchers to share their findings in a more informal, conversational way. These platforms are useful for engaging in discussions with patients and the public.

When communicating research findings, researchers should use clear, concise and accessible language. While a detailed, technical explanation may be necessary for peer-reviewed journals or conference presentations, a simpler, more direct approach is required for the public or policymakers. In these cases, visual aids such as graphs, charts and infographics can be highly effective in illustrating key points and making the data easier to understand. Well-designed visuals help summarise complex information and engage the audience, enhancing the overall impact of the research.

Dr Paul Science Says
'Social media brings both opportunity and responsibility. Use it wisely to share accurate science and engage in meaningful conversations.'

The Challenges of Misinformation

Misinformation is a growing concern in today's digital age, where the rapid spread of false or misleading information can have serious consequences. When communicating research findings, it is important to ensure that the information shared is accurate, reliable and properly contextualised to avoid contributing to the spread of misinformation. One of the challenges researchers face is the potential misinterpretation or oversimplification of their findings by the media or public. Press releases and media coverage can sometimes sensationalise or misrepresent research results, leading to the spread of inaccurate information[20].

Another challenge is the spread of misinformation through social media platforms. The World Health Organization describes this as the 'infodemic'[20]. Researchers who choose to share their findings on social media should be cautious and ensure that their posts are factual, unbiased and supported by evidence. They should be prepared to engage in constructive discussions and address any misconceptions or questions that may arise.

However, they may also face negative experiences, including online 'trolling' and targeted harassment from individuals or groups skeptical of science or opposed to certain research areas. Trolling can range from minor, unconstructive comments to prolonged, intense campaigns designed to undermine credibility, spread misinformation or discourage public engagement. This hostility, often amplified by the anonymity of social media, can make scientists and science communicators feel isolated or even question their commitment to public outreach. Despite the impact of these interactions, many continue their work undeterred. To combat misinformation, researchers can take proactive steps to promote scientific literacy and critical thinking among the public. This can involve taking part in outreach activities, such as public lectures, workshops or science festivals. By meeting the public in person, researchers can engage directly with the audience and explain their research in an accessible manner[20].

> *Dr Paul Science Says*
> 'Clear, honest and accurate communication is our greatest defence against misinformation. Let science be the voice of reason.'

By fostering a better understanding of the scientific process and the importance of evidence-based decision-making, researchers can help build trust in science and counter the spread of misinformation. In addition, researchers should be aware of the potential for their work to be used to support or refute certain political, social or economic agendas. They should strive to present their findings objectively and resist pressure to overstate or understate the implications of their research. When discussing their work in public forums, researchers should be transparent about the limitations, uncertainties and potential biases of their studies and avoid making claims that go beyond the scope of their data. Communicating research findings in the face of misinformation requires researchers to be proactive, cautious and transparent.

Dr Paul Science Says
'It's essential to be transparent about research limitations. Responsible science communication means acknowledging the unknowns.'

Translating Research into Practice

Translating research findings into practice is the process of applying knowledge gained from research studies to real-world settings, such as clinical care, public health or policymaking. This process is often referred to as knowledge translation or implementation science. The first step is to identify the key aspects of the research that have the potential to improve practice. For example, if a study shows that a new medication is highly effective in lowering blood pressure, researchers need to consider how this finding can be applied to patient care. This involves thinking about how the medication can be introduced into routine medical practice and how it might fit into the current treatment options available to healthcare providers. The goal at this stage is to take the knowledge gained from the study and see how it can be used to benefit patients or the public.

Before implementing these findings, researchers must assess whether the new practice is feasible in a real-world setting. Just because a new treatment works well in a controlled research environment doesn't always mean it will be practical to use in everyday healthcare. For example, a new drug might be very effective, but if it is too expensive or has side effects that make patients reluctant to take it, its real-world impact could be limited.

Engaging with the people who will be affected by the new practices is an essential part of the process. This includes healthcare providers, patients and policymakers. By discussing the potential changes with these stakeholders, researchers can gain insights into their needs, concerns and any obstacles they might face. For instance, a doctor might need reassurance that the new medication will be covered by insurance or a patient might want to know more about potential side effects. These conversations are key to making sure that the new practice is not only effective but also widely accepted and adopted.

After the new practices have been implemented, the next step is to evaluate their impact. This means collecting data to see whether the practices are being used as intended and whether they are producing the desired outcomes. In the case of the blood pressure medication, researchers would check if patients' blood pressure were being successfully reduced in regular medical practice, not just in the controlled environment of the original study. They might also look at other factors, such as whether patients are adhering to the medication and whether healthcare providers are prescribing it appropriately.

Translating research into practice is not without its challenges. One of the major difficulties is that moving from the controlled environment of a research study to the complexities of real-world settings can be complicated. For example, a treatment that works well in a large hospital with extensive resources might need to be adapted for use in smaller clinics that have fewer staff or equipment. Resistance to change is a common issue, as healthcare providers may be hesitant to adopt new methods, particularly if they feel under-resourced or overburdened. In these cases, providing additional support may help. This might involve offering workshops, online courses or one-on-one mentoring to help providers integrate the new methods into their daily routines.

The role of Patient and Public Involvement in Shaping Research Priorities

Patient and public involvement (PPI) in research is a collaborative process where researchers work closely with patients, caregivers and members of the community throughout the entire research journey[21]. This approach ensures that the voices of those most affected by the health issues being studied are heard and considered at every stage. Involving patients and the public can significantly enhance the relevance and impact of the research by aligning it more closely with the real-world needs and priorities of the people it aims to benefit.

PPI begins with the very foundation of a research project: identifying the research questions and setting priorities. Often, patients and the public have a different perspective than researchers when it comes to what matters most in healthcare[21]. For instance,

while a researcher might be focused on the bio-
logical mechanisms of a disease, patients may be
more concerned about how a condition affects their
day-to-day lives or how treatments impact their
quality of life. By engaging with patients and the
public early in the process, researchers can ensure
that the questions they are asking address the real
concerns and priorities of those affected.

Beyond shaping research questions, patients and the
public can provide valuable input on the design of
the study itself. They can offer insights into how best
to recruit participants, especially when reaching out
to specific communities or populations.

PPI also plays a key role in the interpretation and dis-
semination of research findings. Patients and community members can help researchers translate complex findings into plain language, ensuring that the
key messages are clear and accessible. They can also
suggest the best ways to share these findings with
their communities, whether through local organisations, public forums or social media.

Another important aspect of PPI is advocating for
the implementation of research findings into practice and policy. Patients and the public are often in
a unique position to champion the use of new treat-

ments, interventions or policies that emerge from research. Their lived experiences can add weight to the call for change. By giving patients and the public an active role in shaping research priorities, researchers can better align their work with the values, goals and concerns of the people they aim to serve. This not only makes the research more patient-centred but also enhances its overall impact.

The benefits of PPI extend beyond the immediate research project. When patients and the public are involved in research, the results are often more applicable, sustainable and likely to drive long-term improvements in health outcomes and quality of life. Involving the public also helps build trust between researchers and the community. It demonstrates a commitment to transparency, collaboration and shared decision-making. This trust is essential for ensuring that future research continues to engage with and benefit the people it is intended to help.

Key Takeaways

- Effective communication ensures research findings reach the healthcare providers, policymakers, patients and the public who can benefit.

- Peer-reviewed journals, conferences, media releases and social media are key channels for sharing research findings.

- Misinformation can distort research findings, so researchers must communicate clearly and factually, to prevent misinterpretations and foster scientific literacy.

- Knowledge translation helps apply research findings in real-world settings, such as clinical care and policy.

Chapter Ten

Future Directions and Challenges in Health Research

Health research is an ever-evolving field, shaped by advances in science, technology and our growing understanding of human health and disease. As new trends and technologies emerge, they are transforming the way research is conducted. These innovations not only enhance the ability to prevent, diagnose and treat diseases, but also allow for more personalised and efficient approaches to healthcare.

One of the most significant emerging trends is precision medicine. This is where prevention, diagnosis and treatment strategies are tailored to an individual's unique genetic makeup, environmental influences and lifestyle choices. Traditional medicine often relies on standard treatments that work for large groups of people, but precision medicine recognises that individuals respond differently to treatments based on their biology and circumstances.

By harnessing big data, genomics and advanced machine learning techniques, researchers can identify personalised risk factors for diseases and create more effective interventions. For example, by analysing a patient's genome, doctors may be able to predict their likelihood of developing certain cancers and recommend tailored screening and prevention strategies. Precision medicine is helping to shift healthcare towards a more customised approach, where treatments are not one-size-fits-all but are specifically designed to meet the needs of each individual.

Dr Paul Science Says
'Precision medicine shows us that one-size-fits-all solutions don't belong in modern healthcare.'

Another transformative trend is the rise of digital health technologies, such as mobile apps, wearable devices and telemedicine platforms. These digital tools are revolutionising how health data are collected, monitored and used for both research and care delivery. Wearable devices, like fitness trackers and smartwatches, can monitor vital signs such as heart rate, activity levels and sleep patterns in real time.

This allows researchers and healthcare providers to gather continuous data on a person's health outside of the clinical setting. In addition, mobile apps can help patients track symptoms, medication adherence and lifestyle behaviours, providing valuable insights into their day-to-day health.

Digital health also enables remote participation in research studies, making it easier for people to contribute to research without having to travel to a clinic. Telemedicine, which allows patients to consult healthcare professionals via video calls, has further expanded access to care, particularly during the COVID-19 pandemic, and is likely to remain a key component of future healthcare delivery.

Artificial intelligence (AI) is another rapidly advancing technology with profound implications for health research. AI and machine learning algorithms can process and analyse vast amounts of complex data more quickly and accurately than humans. This capability is particularly valuable when dealing with large datasets such as electronic health records, medical imaging scans and genomic sequences.

For example, AI can help identify patterns in medical images that may be missed by the human eye, aiding in the early detection of conditions like cancer. Similarly, machine learning algorithms can analyse a patient's medical history and genetic data to predict their risk of developing certain diseases or to recommend the most effective treatments. AI is also being used to support clinical decision-making, providing doctors with data-driven insights to improve patient care. As AI continues to develop, its ability to analyse complex health data will enhance researchers' understanding of diseases and improve outcomes in clinical practice.

Dr Paul Science Says
'AI is our most powerful research partner yet, capable of finding patterns and insights in data that were once invisible to the human eye.'

Regenerative medicine is another exciting area that has the potential to change the landscape of healthcare. This field focuses on developing therapies that can repair, replace or regenerate damaged tissues and organs, offering hope for people with chronic or degenerative conditions. One of the most promising approaches in regenerative medicine is the use of stem cells, which can develop into many different types of cells in the body.

Stem cell therapies are being explored for a range of conditions, from heart disease to spinal cord injuries, with the aim of regenerating damaged tissues and restoring function.

Tissue engineering is another aspect of regenerative medicine, where researchers are developing lab-grown tissues and organs that could one day be used for transplants.

Also, gene editing techniques such as CRISPR are being studied as potential treatments for genetic disorders, allowing scientists to correct faulty genes that cause disease. CRISPR is a tool scientists use to change DNA, which is the code inside our cells that determines things like our traits and how our bodies work.

It acts like molecular scissors, allowing scientists to cut specific parts of DNA and either remove, add or change bits of genetic information. This technology has huge potential for treating diseases, improving crops and even studying how genes affect health. Essentially, CRISPR makes it much easier and faster to edit genes than older methods.

Dr Paul Science Says
'The convergence of science and technology is rewriting the rules of healthcare. Precision medicine, AI, and regenerative therapies are just the beginning.'

Key Takeaways

- Precision medicine is shifting healthcare from one-size-fits-all treatments to personalised approaches.

- Digital health tools, such as mobile apps, wearable devices and telemedicine, are revolutionising data collection

- AI and machine learning are transforming health research by analysing complex data more quickly and accurately than humans.

- Regenerative medicine, including stem cell therapies and gene editing, is creating new possibilities for repairing and regenerating damaged tissues and organs.

As these emerging trends and technologies continue
to evolve, they are likely to reshape the future of
health research in profound ways. Precision medicine is leading the charge towards more personalised healthcare, digital health is expanding access
and improving data collection, AI is enhancing the
ability to analyse complex health information and
regenerative medicine is offering new avenues for
treatment and recovery.

Dr Paul Science Says
'As health researchers, our job is to turn today's
innovations into tomorrow's healthcare stan-
dards, ensuring they are safe, effective and accessible to everyone.'

Together, these innovations are paving the way for more effective, efficient and patient-centred approaches to healthcare. The potential to prevent, diagnose and treat diseases in ways that were previously unimaginable is becoming a reality.

In the years to come, these rapid technological advancements and scientific discoveries will not only transform the way research is conducted but will also have a lasting impact on how healthcare is delivered. These changes will ultimately improve health outcomes and quality of life for patients around the world.

As we embrace these new technologies, we must ensure that they are used in ways that benefit all people. This is to ensure that we foster in an era of healthcare that is not only more advanced but also more equitable and accessible.

About the Author

PAUL IAN CROSS, BSc (Hons), MSc, PhD, is a scientist, health researcher and speculative fiction author.

With over twenty years' experience in clinical research, Paul has been involved in the design, set-up or delivery of over a hundred clinical studies, including first-in-human clinical trials in gene and cell therapy and COVID-19 vaccine research.

Paul provides consultancy services to organisations who oversee or manage clinical research, including clinical trial set-up, ICH Good Clinical Practice oversight and project management, science communication, science writing and research training.

With a keen focus on education, Paul strives to make STEM subjects – Science, Technology, Engineering and Mathematics – accessible and interesting for both children and adults. This educational drive is evident in the compelling STEM content he shares across his social media platforms.

He is an award–nominated author whose work spans multiple genres, including STEM-inspired books for children and young adults and speculative fiction for everyone else.

With a solid foundation in biology and a keen interest in health, medicine and nature, Paul's scientific curiosity also spans technology, environmental sustainability, climate change and space exploration. His fascination with artificial intelligence and the possibility of alien life often finds its way into his science fiction narratives, adding a layer of intrigue to his stories.

Paul's aim is not only to inform but to inspire, making science accessible and fun for a wider audience.

Connect with Paul
Ian Cross

pauliancross.com

Bluesky: @pauliancross.bsky.social

Facebook: /pauliancrossauthor

Instagram: @DrPaulScience

LinkedIn: @pauliancross

Threads: @DrPaulScience

TikTok: @DrPaulScience

YouTube: @DrPaulScience

X (Twitter): @pauliancross

Please leave a review for this edition of *Health Research Explainers*.

Thank you.

References

(1) National Institute for Health and Care Research. Health and care research: Introduction [Internet]. 2024 [cited 2024 Oct 16]. Available from: https://www.nihr.ac.uk/career-development/health-and-care-research-introduction

(2) World Health Organization. Social determinants of health [Internet]. 2024 [cited 2024 Oct 16]. Available from: https://www.who.int/health-topics/social-determinants-of-health#tab=tab_1

(3) Mahmood SS, Levy D, Vasan RS, Wang TJ. The Framingham Heart Study and the epidemiology of cardiovascular diseases: a historical perspective. *Lancet*. 2014 Mar 15;383(9921):999–1008.

(4) BBC Bitesize. The Human Genome Project and medicine [Internet]. 2023 [cited 2024 Oct 16]. Available from: https://www.bbc.co.uk/bitesize/guides/z8qcsrd/revision/10

(5) World Health Organization. Vaccines and immunization: What is vaccination? [Internet]. 2020 Dec 30 [cited 2024 Oct 19]. Available from: https://www.who.int/news-room/questions-and-answers/item/vaccines-and-immunization-what-is-vaccination

(6) Pfizer. About our landmark trial [Internet]. 2021 [cited 2024 Oct 16]. Available from: https://www.pfizer.com/science/coronavirus/vaccine/about-our-landmark-trial

(7) World Health Organization. Who can take the Pfizer-BioNTech COVID-19 vaccine: What you need to know [Internet]. 2021 [cited 2024 Oct 16]. Available from: https://www.who.int/news-room/feature-stories/detail/who-can-take-the-pfizer-biontech-covid-19--vaccine-what-you-need-to-know

(8) American Museum of Natural History. The Scientific Process [Internet]. Available from: https://www.amnh.org/explore/videos/the-scientific-process

(9) ICH. ICH E6(R3) draft guideline. [Internet]. 2023 May 19 [cited 25th April 2024]. Available from: https://ich.org/page/efficacy-guidelines/

(10) The Nuremberg Code. BMJ [Internet]. 1996 [cited 2024 Oct 24];313(7070):1448.Available: https://media.tghn.org/medialibrary/2011/04/BMJ_ No_7070_Volume_313_The_Nuremberg_Code.pd f

(11) World Medical Association. WMA Declaration of Helsinki – Ethical Principles for Medical Research Involving Human Subjects [Internet]. Ferney-Voltaire: World Medical Association; 2013 [cited 2024 Oct 24]. Available from: https://www.wma.net/what-we-do/medical -ethics/declaration-of-helsinki/

(12) Tuskegee University. About the USPHS Syphilis Study [Internet]. Tuskegee: Tuskegee University; [cited 2024 Oct 24]. Available from: https://www.tuskegee.edu/about-us/centers-of-exc ellence/bioethics-center/about-the-usphs-syphilis-s tudy

(13) U.S. Department of Health and Human Services. The Belmont Report: Ethical Principles and Guidelines for the Protection of Human Subjects of Research [Internet]. Washington, DC: U.S. Department of Health and Human Services; 1979 [cited 2024 Oct 24]. Avail-

able from: https://www.hhs.gov/ohrp/regulations -and-policy/belmont-report/index.html

(14) The Open University. Understanding different research perspectives. OpenLearn. Available from: https://www.open.edu/openlearn/money-business/ understanding-different-research-perspectives/cont ent-section-5

(15) Chidambaram AG, Josephson M. Clinical research study designs: The essentials. Pediatr Investig [Internet]. 2019 Dec;3(4):245–252. Published online 2019 Dec 21. doi: 10.1002/ped4.12166. PM-CID: PMC7331444. PMID: 32851330. Available from: https://www.ncbi.nlm.nih.gov/pmc/articles/ PMC7331444/

(16) National Institute for Health and Care Research. What is health and care research? Be Part of Research. Available from: https://bepartofresearch.nih r.ac.uk/what-is-health-and-care-research/

(17) Elfil M, Negida A. Sampling methods in Clinical Research; an Educational Review. Emerg (Tehran). 2017;5(1)

(18) Martínez-Mesa J, González-Chica DA, Bastos JL, Bonamigo RR, Duquia RP. Sample size: how many participants do I need in my research? An

Bras Dermatol. 2014 Jul-Aug;89(4):609-15. doi: 1 0.1590/abd1806-4841.20143705. PMID: 25054748; PMCID: PMC4148275.

(19) Sardana N, Shekoohi S, Cornett EM, Kaye AD. Chapter 6 - Qualitative and quantitative research methods. In: Kaye AD, Urman RD, Cornett EM, Edinoff AN, editors. Substance Use and Addiction Research. Academic Press; 2023. p. 65-9. ISBN: 9780323988148.

(20) World Health Organization. Understanding the infodemic and misinformation in the fight against COVID-19 [Internet]. Geneva: World Health Organization; 2020 [cited 2024 Oct 26]. Available from: https://www.who.int/health-topics/infodemic/unde rstanding-the-infodemic-and-misinformation-in-t he-fight-against-covid-19#tab=tab_1

(21) National Institute for Health Research. PPI (Patient and Public Involvement) resources for applicants to NIHR research programmes [Internet]. London: NIHR; 2019 Dec 18 [cited 26th April 2024]. Available from: https://www.nihr.ac.uk/documents/ppi-patient-an d-public-involvement-resources-for-applicants-to -nihr-research-programmes/23437